TEACHING FRACTIONS
USING
LEGO® BRICKS

Dr. Shirley Disseler

COMPASS

Teaching Fractions Using LEGO® Bricks

Copyright ©2016 by Shirley Disseler
Published by Brigantine Media/Compass Publishing
211 North Avenue, St. Johnsbury, Vermont 05819

Cover and book design by Anne LoCascio
Illustrations by Curt Spannraft

Brigantine Media/Compass Publishing
211 North Avenue
St. Johnsbury, Vermont 05819
Phone: 802-751-8802
Fax: 802-751-8804
E-mail: neil@brigantinemedia.com
Website: www.compasspublishing.org

ORDERING INFORMATION
Quantity sales
Special discounts for schools are available for quantity purchases of physical books and digital downloads.
For information, contact Brigantine Media at the address shown above or visit
www.compasspublishing.org.

Individual sales
Brigantine Media/Compass Publishing publications are available through most booksellers.
They can also be ordered directly from the publisher.
Phone: 802-751-8802 | Fax: 802-751-8804
www.compasspublishing.org
ISBN 978-1-9384065-6-0

CONTENTS

DEDICATION

To my sons Steven and Ryan, whose love of LEGO® bricks as children inspired me to find ways to use the bricks in education to engage young minds in math!

INTRODUCTION

Fractions! This word can bring students and parents to tears during the school year. Many students never fully understand the meaning of the word *fraction*.

The term *fraction* refers to a whole divided into parts, but the problem lies in understanding how to define the term *whole*. Wholes are different, making fractional amounts different. For example: $\frac{1}{2}$ of a whole carton of eggs is 6 eggs if the carton holds a true dozen; however, $\frac{1}{2}$ of a carton that holds 18 eggs is 9.

Students also have trouble understanding that a fraction represents just one number, despite the fact that a fraction is made of two digits, a numerator and a denominator.

Understanding fractions relies on a student's knowledge of multiplication tables, which can also be an issue. To find common denominators, students must know how to find factors, and therefore, know multiplication tables.

By the end of elementary school, students should have learned:
- how to recognize fractions
- the vocabulary of fractions
- how to define a whole and represent parts of different sized wholes
- how to add and subtract fractions with like and unlike denominators
- how to represent and understand mixed numbers and find equivalent fractions

This book will help students master these topics, using a material found in almost every classroom and home—LEGO® bricks.

Why use LEGO® bricks to learn about fractions?

LEGO® bricks help students learn mathematical concepts through modeling. If a student can model a math problem, and then be able to understand and explain the model, he or she will begin the computational process without struggling.

Modeling fractions with LEGO® bricks is an easy way for students to demonstrate understanding of the vocabulary and the concepts of fractional numbers. When students model fractions with LEGO® bricks, they have the opportunity to create multiple solutions for problems instead of looking for only one right answer.

LEGO® bricks are great tools for bringing many mathematical concepts to life: basic cardinality and counting, addition and subtraction, multiplication and division, fractions, data and measurement, and statistics and probability. Using LEGO® bricks fosters discussion, modeling, collaboration, and problem solving. These are the 21st century skills that will help our students learn and be globally competitive.

The use of a common child's toy to do math provides a universal language for math. Children everywhere recognize this manipulative. It's fun to learn when you're using LEGO® bricks!

USING A BRICK MATH JOURNAL

Journaling in math is an exciting way for students, teachers, and parents to review and share what is going on in the math classroom. A math journal is a resource that students can use for years to practice and review math concepts.

I recommend having your students start a Brick Math journal when you begin using LEGO® bricks to help teach math concepts. Here's how to use a Brick Math journal with the activities in this book: In each chapter, students begin in Part 1 (Show Them How) by building models that are teacher-directed. In Part 2 (Show What You Know), students build their own LEGO® brick models in response to specific prompts. Finally, students draw their LEGO® brick models on paper in their own Brick Math journals. The journal serves as a record of the physical models built that students can refer to over and over. The Brick Math journal can also serve as a form of assessment for teachers, a source for conferences, and as a way to identify if a student has any misconceptions in learning the topic.

Use these steps to create a useful Brick Math journal:

1. Use a composition book for each student. Set up pages one and two as the table of contents.

2. Have students number each page on the top outside corner.

3. Photocopy the base plate paper in the Appendix. Students will use the base plate paper to record their solutions, drawings, reflections, etc. Students will glue the base plate paper onto a journal page after they have colored/sketched solutions.

4. When students glue in their drawings, they should label them with the title of the activity, and then make the entry in the journal's table of contents.

.

SUGGESTED BRICKS

Size	Number
1x1	1-64
1x2	1-32
2x2	2-16
	(various colors)
2x4	10-12
	(two colors)

Note: Using a base plate will help keep the bricks in a uniform line. One base plate is suggested for these activities.

PARTS OF A FRACTION

Students will learn/discover:
- The term *numerator*
- The term *denominator*
- The term *whole* and how it is represented in a fraction

Why is this important?
Students need to see a fraction as one number, not as two separate numbers.

They also need to understand the relationship between the numerator and denominator as they proceed toward ratios in later content.

Brick Math journal:
After students build their models, have them draw the models on base plate paper and keep them in their Brick Math journals (see page 7 for instructions). Recording the models on paper after building with the LEGO® bricks helps to reinforce the concepts.

Part 1: Show Them How
Model parts of a fraction

1. A fraction represents part of a whole.

A fraction has two parts: the *numerator* and the *denominator*. The denominator is the digit that tells the size of the whole. For example, if the denominator is 8, then we know that the whole is made up of 8 pieces. The denominator is located on the bottom of a fraction. The numerator is the digit that tells the amount being used out of the whole. For example, if the numerator is 6 and the denominator is 8, then we know that we are using 6 of the 8 pieces or $\frac{6}{8}$.

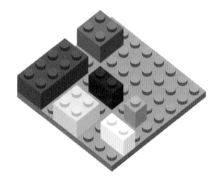

2. Use the LEGO® bricks to make models of ½.

Ask students to think about what the term "half" means.

When using the LEGO® bricks to model fractions, count the studs. The studs represent the digits of the denominator and numerator of the fraction. We can model ½ in many ways.

3. Ask students: Which colors of LEGO® bricks represent numerators?

(*Answer:* blue 2x2, red 1x2, and orange 1x1)

4. Ask students: Which colors represent denominators?

(*Answer:* green 2x4, yellow 2x2, and white 1x2)

5. Ask students: Can you show two other ways to make ½?

Part 2: Show What You Know #1

1. Can you design a model that shows ¼?

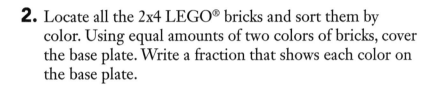

Show What You Know #2

1. Take a base plate. Count the number of studs on your base plate.

(*Answer:* 64 studs)

2. Locate all the 2x4 LEGO® bricks and sort them by color. Using equal amounts of two colors of bricks, cover the base plate. Write a fraction that shows each color on the base plate.

(*Answer:* There are a number of solutions, but the idea is to model each half of the base plate in a different color. Students will describe this model in different ways depending on their level of understanding fractions. Some will answer "$^{32}/_{64}$ yellow and $^{32}/_{64}$ red," if they are counting studs. Some will see that they have used eight bricks in total and answer "$^{4}/_{8}$ yellow and $^{4}/_{8}$ red." With this model the most appropriate answer is $^{4}/_{8} = ^{1}/_{2}$. Make sure students recognize that their models show two halves.

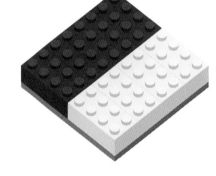

3. Using the same size base plate, cover the base plate with single 1x1 bricks of two different colors. How is this model different from the model just created?

(*Answer:* Students will respond differently depending on their level of understanding fractions. What's important here is that they give a different answer than for the previous model. Students need to understand that this model shows a different whole. This whole has 64 individual parts or 64 studs.

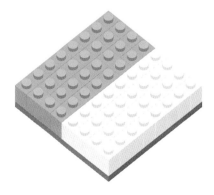

They could describe the model as: "half of the model is orange and half is white." An alternative correct answer is "the orange is $^{32}/_{64}$ and the white is $^{32}/_{64}$.")

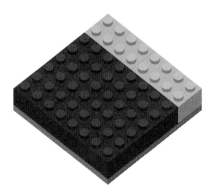

$^{16}/_{64}$, $^2/_8$, or $^1/_4$

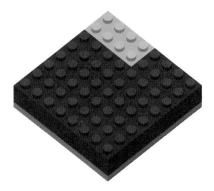

$^8/_{64}$ or $^1/_8$

4. Can you use this base plate and other colors of bricks to create a model for ¼? ⅛? Sketch your solutions in your Brick Math journal.

(Answer: A variety of solutions are acceptable. In these models, students are putting together the idea of modeling eight parts as the whole with the idea of counting studs using eight bricks comprised of 64 studs.)

SUGGESTED BRICKS

Size	Number
1x1	4
1x2	6-8
1x4	4-6
2x2	4-6
2x4	9-12
2x8	2

Note: Using a base plate will help keep the bricks in a uniform line. One base plate is suggested for these activities.

BENCHMARK FRACTIONS

Students will learn/discover:
- Values of the benchmark fractions: $\frac{1}{4}$, $\frac{1}{2}$, and $\frac{3}{4}$
- Fractions can be made with different wholes

Why is this important?
Students start by understanding the simple benchmark fractions that they will see over and over again. Benchmark fractions better prepare students to make estimates of measurements, distances, and amounts of wholes in real-life situations.

Students must also understand that fractions can be made from many different wholes.

Definition: Different wholes
All wholes are not equal. One-eighth of a 14-inch large pizza is not the same size piece as one-eighth of an 8-inch small pizza. The fractional amount may be the same, but if the fraction comes from different sized wholes, the fractions are not equal.

Brick Math journal:
After students build their models, have them draw the models on base plate paper and keep them in their Brick Math journals (see page 7 for instructions). Recording the models on paper after building with the LEGO® bricks helps to reinforce the concepts.

Part 1: Show Them How
Model benchmark fractions $\frac{1}{2}$, $\frac{1}{4}$, $\frac{1}{8}$, $\frac{3}{4}$

1. Place the 2x4 brick on the base plate. This is the whole. Remind students that this represents the "denominator" of the fraction. The denominator is the whole that is being divided into parts. It is the bottom number in a fraction.

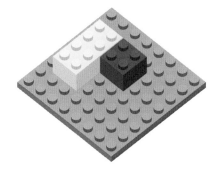

2. Find $\frac{1}{2}$ of this whole. Look for two bricks that are the same size that take up the same space as the whole when they are placed together. One of these bricks equals $\frac{1}{2}$.

(*Answer:* one 2x2 brick)

The four-stud brick (2x2) is one-half the eight-stud brick (2x4). The 4 is the "numerator" in the fraction. In a fraction this would be the top number. Since 4 is half of 8, this fraction can be written as $\frac{4}{8}$ or $\frac{1}{2}$.

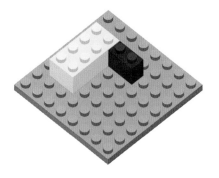

3. Find $\frac{1}{4}$ of the whole. Look for four bricks that are the same size that take up the same space as the whole when placed together. One of these bricks equals $\frac{1}{4}$.

(*Answer:* one 1x2 brick)

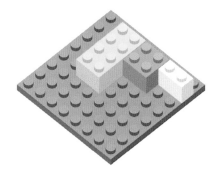

More for students to discover:
Demonstrate that $\frac{1}{4}$ is equivalent to $\frac{1}{2}$ of the half found in step 2.

4. Find ¹/₈ of the whole. Look for eight equal-sized bricks that take up the same space as the whole when placed together. One of these bricks equals ¹/₈.

(*Answer:* one 1x1 brick)

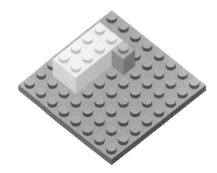

More for students to discover:
Demonstrate that this 1x1 brick is equivalent to ¹/₂ of the ¹/₄ in step 3.

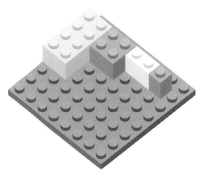

5. Find ³/₄ of the whole. Look for the brick that made ¹/₄, then find three of them.

(*Answer:* three 1x2 bricks)

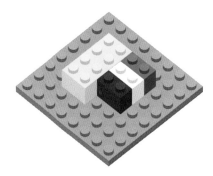

More for students to discover:
Demonstrate that three 1x2 bricks are equivalent to one 2x3 brick.

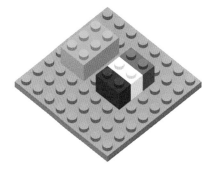

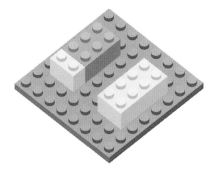

Even more for students to discover:
Demonstrate how to add ¾ and ¼ to get ⁴⁄₄, which is equivalent to 1 whole.

Part 2: Show What You Know

1. If one whole is equal to the 2x8 brick, what bricks can make ½ of that whole?

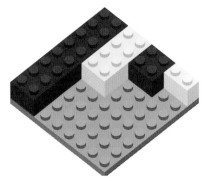

2. What bricks make ¼ of that whole?

3. What bricks make ⅛ of that whole?

The whole, ½, ¼, ⅛

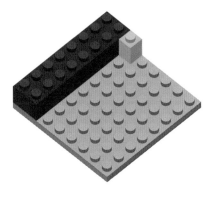

4. What bricks make ¹⁄₁₆ of that whole?

The whole and ¹⁄₁₆

5. What bricks make ¾ of that whole?

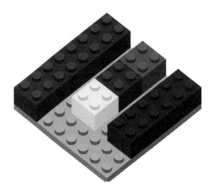

The whole and ³/₄
(three of the 2x2 bricks OR
one 2x6 brick)

6. Show how to add ¼ and ¾

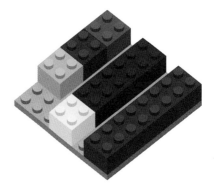

left: three ¹/₄ portions of the whole
middle: ³/₄ + ¹/₄
right: 1 whole

3

SUGGESTED BRICKS

Size	Number
1x1	20
1x2	8
1x3	8
2x3	8
2x4	8
2x8	4
	(can use eight 2x4 bricks if necessary)

Note: Using a base plate will help keep the bricks in a uniform line. Use two small base plates or one large base plate for these activities.

ADDING FRACTIONS WITH LIKE DENOMINATORS

Students will learn/discover:
- How to add two fractions whose denominators are the same

Why is this important?
Students who can recognize parts of a whole are ready to learn how to add parts of the same whole.

Note: Students should not add with unlike denominators in this activity.

Brick Math journal:
After students build their models, have them draw the models on base plate paper and keep them in their Brick Math journals (see page 7 for instructions). Recording the models on paper after building with the LEGO® bricks helps to reinforce the concepts.

Part 1: Show Them How
Model adding two fractions: ³⁄₈ + ²⁄₈

1. Place two 2x4 bricks on the base plate. Use the same color 2x4 bricks to help students understand the idea of *like denominators*.

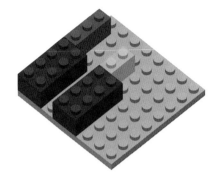

2. Above one 2x4 brick, place one 1x3 brick. Above the other 2x4 brick, place one 1x2 brick. Use different color 1x3 and 1x2 bricks to demonstrate that the numerators are not the same.

3. To model the solution to the problem ³⁄₈ + ²⁄₈ = ⁵⁄₈, use a second base plate and move the bricks over from your model.

Place the 2x4 (eight-stud) brick at the bottom of the base plate. Place the 1x3 brick above the 2x4 brick. Place the 1x2 brick above the 1x3 brick. Place a 1x1 brick on top of each stud on the 1x3 and 1x2 bricks to show that five 1x1 bricks are equivalent to the 1x3 and 1x2 bricks added together.

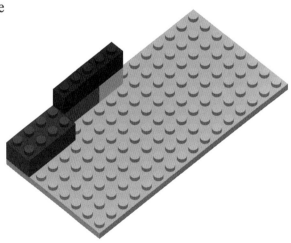

Note: This model requires one base plate that is larger than 8x8 or two 8x8 base plates.

Note: There are no 1x5 bricks. Keep this in mind when planning problems.

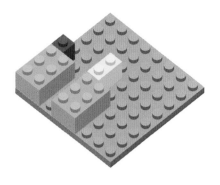

Part 2: Show What You Know #1
Adding two fractions with like denominators

1. Can you model $\frac{1}{6} + \frac{2}{6}$?

(*Answer:* Use a 2x3 brick for each denominator. Place a 1x1 brick above the first denominator brick, and a 1x2 brick above the second denominator brick.)

Note: Check to make sure students use the same color bricks to model the denominators.

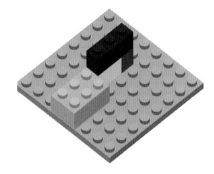

2. Model the solution to $\frac{1}{6} + \frac{2}{6}$.

(*Answer:* Move the bricks from the first model to a new base plate to show the solution as $\frac{3}{6}$, using the same technique as in Part 1 to construct a model that shows $\frac{1}{6} + \frac{2}{6} = \frac{3}{6}$.)

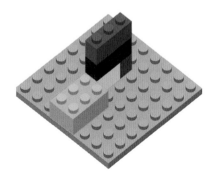

More for students to discover:
Can you show that $\frac{3}{6} = \frac{1}{2}$?

(*Answer:* place a 1x3 brick on top of the three 1x1 bricks. The 1x3 brick is $\frac{1}{2}$ the 2x3 brick.)

Show What You Know #2
Adding three fractions with like denominators

1. Can you model $^6/_{14}$ + $^4/_{14}$ + $^2/_{14}$?

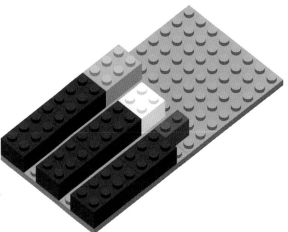

(*Answer:* Make three stacks, each with one 2x4 and one 2x3 brick of the same color to represent the denominator of 14. Place a 2x3 brick above one denominator, a 2x2 brick above the second denominator, and one 1x2 brick above the third. Make sure each of the numerator bricks is a different color.)

2. Model the solution to $^6/_{14}$ + $^4/_{14}$ + $^2/_{14}$.

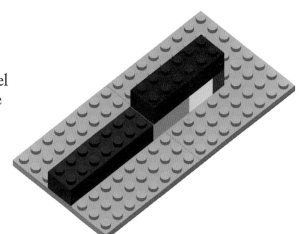

(*Answer:* Move the fraction to a new base plate to model the solution. Use the same technique as earlier to place all the numerator bricks above the denominator brick, then stack 1x1 bricks on each stud of the numerator bricks and count them to find the answer:
$^6/_{14}$ + $^4/_{14}$ + $^2/_{14}$ = $^{12}/_{14}$)

Additional exercise:
Have students explain their models in writing.

Note: You'll learn a lot from having the students write the rationale for their models, and you can help students at any step where they aren't clear about how to add fractions with like denominators.

4

SUGGESTED BRICKS

Size	Number
1x1	10
1x2	8
1x3	4-8
2x2	5-8
2x3	8
2x4	6-8
2x6	6-8

Note: Using a base plate will help keep the bricks in a uniform line. Two base plates are suggested for these activities.

SUBTRACTING FRACTIONS WITH LIKE DENOMINATORS

Students will learn/discover:

*   How to subtract two fractions whose denominators are the same

Why is this important?

Students who can recognize parts of a whole are ready to learn how to subtract parts of the same whole.

Note: Students should not subtract with unlike denominators in this activity.

This activity can follow learning to add fractions with like denominators or be done at the same time.

Brick Math journal:

After students build their models, have them draw the models on base plate paper and keep them in their Brick Math journals (see page 7 for instructions). Recording the models on paper after building with the LEGO® bricks helps to reinforce the concepts.

Part 1: Show Them How
Model subtracting two fractions: $^3/_8$ - $^1/_8$

1. Place two 2x4 bricks on the base plate. Use the same color and shape 2x4 bricks to help students understand the idea of *like denominators*.

2. Above one 2x4 brick, place one 1x3 brick. Place it horizontally, rather than vertically as when modeling addition. Above the other 2x4 brick, place one 1x1 brick. Use different color 1x3 and 1x1 bricks to demonstrate that these numerators are not the same.

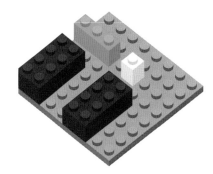

3. To model the solution to the problem, use another base plate and use more bricks to model. Place one 2x4 brick at the bottom of the base plate. Place a 1x3 brick above the 2x4 brick, aligning it horizontally. This is the numerator of the minuend.

Stack the 1x1 brick on top of the 1x3 brick. This is the numerator of the subtrahend.

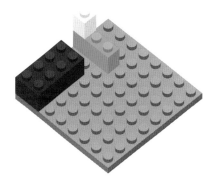

4. Count the studs not covered to show that two studs are left after subtracting the 1x1 brick from the 1x3 brick. This demonstrates the solution to $^3/_8$ - $^1/_8$ = $^2/_8$.

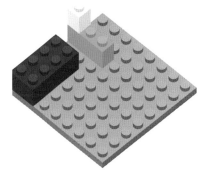

Part 2: Show What You Know
Subtracting two fractions with like denominators

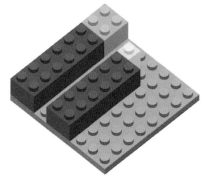

1. Can you model $^4/_{12} - {}^1/_{12}$?

(*Answer:* Use a combination of bricks that total twelve studs to model both denominators. Students could use one 2x6 brick, two 1x6 bricks, two 2x3 bricks, or one 2x4 brick and one 2x2 brick. Make sure all the bricks used for both denominators are the same color, to demonstrate that they are like denominators.

Stack a 2x2 brick (or a 1x4 brick) above one denominator and a 1x1 brick on top of the other denominator.)

More for students to discover:
Using the terms *minuend* and *subtrahend*

Have students identify that the $^4/_{12}$ brick model represents the minuend and the $^1/_{12}$ brick model represents the subtrahend.

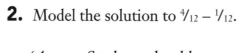

2. Model the solution to $^4/_{12} - {}^1/_{12}$.

(*Answer:* Students should use a new base plate and more bricks to show the solution as $^3/_{12}$. They should use the same technique as in Part 1 to construct a model showing $^4/_{12} - {}^1/_{12} = {}^3/_{12}$.)

More for students to discover:
For students who are ready, ask them to show the solution in simplest form.

(Answer: Find out how many 1x3 bricks (numerator) fit onto the 2x6 brick (denominator). Since four 1x3 bricks fit, the simplified fraction is $\frac{1}{4}$.

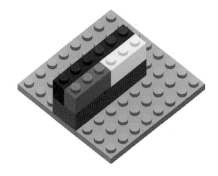

5

SUGGESTED BRICKS

Size	Number
1x1	20
1x2	6-8
1x4	4-6
1x16	2
2x2	4-6
2x4	9-12
2x8	2

Note: Using a base plate will help keep the bricks in a uniform line. One base plate is suggested for these activities.

FACTORS

Students will learn/discover:
- What factors are
- How to find all the factors of numbers
- How to make models of factor families

Why is this important?
Students need to be able to identify all the factors of numbers before they can work on equivalent fractions, simplifying fractions, and addition or subtraction of unlike denominators. For example: adding and subtracting fractions with unlike denominators requires a common denominator. Finding a common denominator requires knowing factors.

Definition: Factors
Factors are numbers you can multiply together to get another number. Example: 2 and 3 are factors of 6; 2 and 4 are factors of 8.

Brick Math journal:
After students build their models, have them draw the models on base plate paper and keep them in their Brick Math journals (see page 7 for instructions). Recording the models on paper after building with the LEGO® bricks helps to reinforce the concepts.

Part 1: Show Them How
Model how to find all the factors of 16

1. Place a 2x8 brick or a 1x16 brick on a base plate.

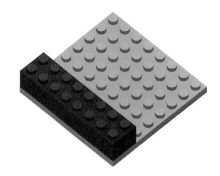

2. Place two bricks that are the same and, when placed next to the 16-stud brick, are equivalent in size and show two halves of the 16-stud brick. Use two 2x4 bricks or two 1x8 bricks.

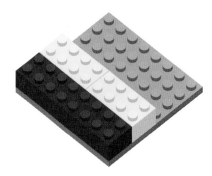

3. Ask students: Can you find three bricks of equal size equivalent to the size of the 16-stud brick?

Let students look and think, and discover that the answer is no.

4. Ask students: Can you find four bricks of equal size equivalent to the size of the 16-stud brick?

Let students look and think, and discover that the answer is four 2x2 bricks or four 1x4 bricks.

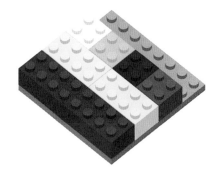

5. Ask students: Can you find the next number of equal-sized bricks that are equivalent to the size of the 16-stud brick?

Let students discover that five, six, and seven bricks don't work. Let them discover that the answer is eight 1x2 bricks.

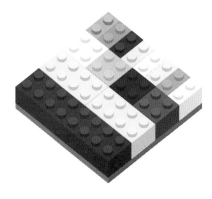

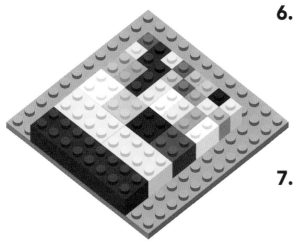

6. Ask students: Can you find the next number of equal-sized bricks that are equivalent to the size of the 16-stud brick?

Let students discover that the answer is sixteen 1x1 bricks.

7. Name all the factors of 16 by looking at the LEGO® bricks on the base plate.

(*Answer:* 16, 8, 4, 2, and 1.)

Part 2: Show What You Know

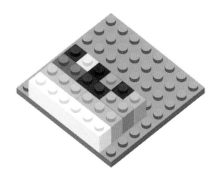

1. Can you build a model to show all the factors of 6?

Solution A:
This single-stud model is a possible solution, showing factors 6, 3, 2, and 1.

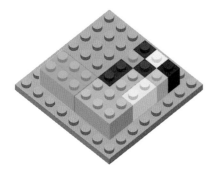

Solution B:
This model shows a combination of single-stud bricks and double-stud bricks. Students who create this model could also explain that there are 2 sets of 3 in 6, and 3 sets of 2 in 6.

Show What You Know

2. Can you build a model to show all
the factors of 8?

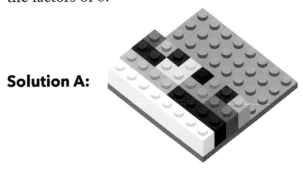

Solution A:

Solution B:

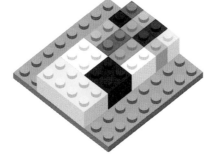

Solution C:

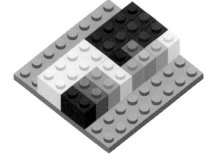

Solutions B and C use a combination of single-stud
bricks and double-stud bricks.

Students who create these models could also explain that
there are 2 sets of 4 in 8, and 4 sets of 2 in 8.

6

SUGGESTED BRICKS

Size	Number
1x1	16 - 20
1x2	12 - 16
1x16	1
2x2	6 - 10
2x4	4 - 6
2x8	2

Note: Using a base plate will help keep the bricks in a uniform line. One base plate is suggested for these activities.

EQUIVALENT FRACTIONS

Students will learn/discover:
- How to set up an equivalent fraction family

Why is this important?
To learn to simplify fractions, students need to be able to recognize equivalent fractions, especially halves, thirds, fourths, and eighths. Finding equivalent fractions is also necessary when adding and subtracting with unlike denominators.

Brick Math journal:
After students build their models, have them draw the models on base plate paper and keep them in their Lego Brick journals (see page 7 for instructions). Recording the models on paper after building with the LEGO® bricks helps to reinforce the concepts.

Part 1: Show Them How
Build a sixteenth fraction model

This is a good model to begin with because it is a large number with many equivalent fractions.

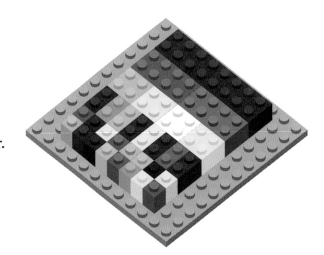

Definition of equivalent fraction: Equivalent fractions are different fractional representations for the same number.

one 16-stud brick (2x8)
two 8-stud bricks (2x4) = halves
four 4-stud bricks (2x2) = fourths
eight 2-stud bricks (1x2) = eighths
sixteen 1 stud bricks (1x1) = sixteenths

Another sixteenth fraction model could be made with:
one 16-stud (1x16) brick
two 8-stud (1x8) bricks
four 4-stud (1x4) bricks
eight 2-stud (1x2) bricks
sixteen 1-stud (1x1) bricks

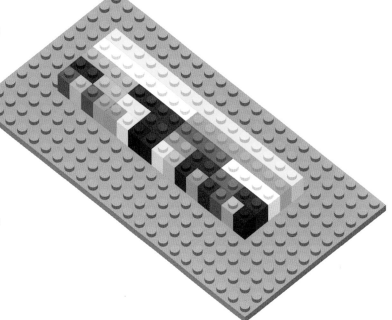

Discuss with students what each brick size represents. Discuss how the model shows:

$^2/_{16} = {}^1/_8$

$^4/_{16} = {}^1/_4$

$^8/_{16} = {}^4/_8 = {}^1/_2$

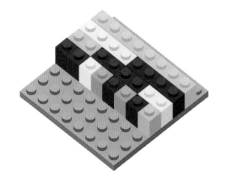

Part 2: Show What You Know

1. Can you make a model of twelfths?

(Answer:
One 1x12 brick as the whole
Divided into two pieces (halves) with two 1x6 bricks
Divided into four pieces (fourths) with four 1x4 bricks
Divided into three pieces (thirds) with four 1x3 bricks
Divided into six pieces (sixths) with six 1x2 bricks
Divided into twelve pieces (twelfths) with twelve 1x1 bricks)

2. Can you make a model of eighths?

(Answer:
One 1x8 brick as the whole
Divided into two pieces (halves) with two 1x4 bricks
Divided into four pieces (fourths) with four 1x2 bricks
Divided into eight pieces (eighths) with eight 1x1 bricks)

3. Choose another whole and make a model to show halves, thirds (if possible), fourths, and eighths.

(Answers will vary)

Try whole numbers such as 18, 24, and 28.

Note: To make wholes greater than 16, combine 2 or more bricks. For example, to make a whole of 24, use two 12-stud bricks (2x6) of the same color placed side by side. To show halves, use two 12-stud bricks in different colors.

ADDING FRACTIONS WITH UNLIKE DENOMINATORS

SUGGESTED BRICKS

Size	Number
1x1	15-20
1x2	12-15
1x3	15
1x4	5
2x2	10
2x3	8
2x4	8
2x8	4
	(or eight 2x4 bricks if necessary)

Note: Using a base plate will help keep the bricks in a uniform line. Use two small base plates or one large base plate for these activities.

Students will learn/discover:
- How to add two fractions whose denominators are the not the same

Why is this important?

Once students have a firm understanding of how to simplify and find equivalents for fractions, they are ready to add fractions of varying sizes. Using common denominators is one of the big ideas of math that prepares students for later mathematical proficiency. We use common denominators to add fractions in many everyday situations. For example: John and Mike ran two legs of a race. John ran $\frac{1}{2}$ of a mile and Mike ran $\frac{2}{3}$ of a mile. How far did the two run altogether? This problem requires a common denominator to add the two distances. One could do a much more involved computation and use measurement and division to find the answer, but being able to find a common denominator and add makes the problem simpler to manage.

Brick Math journal:

After students build their models, have them draw the models on base plate paper and keep them in their Brick Math journals (see page 7 for instructions). Recording the models on paper after building with the LEGO® bricks helps to reinforce the concepts.

Part 1: Show Them How
Model adding two fractions: $\frac{1}{4} + \frac{2}{3}$

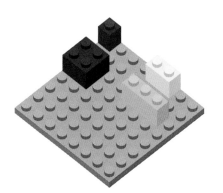

1. Place one 2x2 brick on the base plate with one 1x1 brick above it to model $\frac{1}{4}$. Place one 1x3 brick on the base plate with one 1x2 brick above it to model $\frac{2}{3}$.

Note: Use different colors of LEGO® bricks for numerator and denominator of both fractions to help demonstrate that they are different.

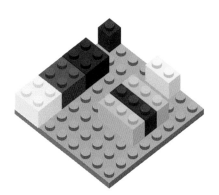

2. Since both denominators are different, add bricks to each denominator until both denominators have an equal number of studs.

Two bricks have been added to each denominator. The $\frac{1}{4}$ fraction's denominator now has 12 studs, but the $\frac{2}{3}$ fraction's denominator has only 9.

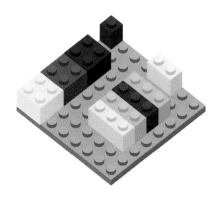

Add one more 1x3 brick to the $\frac{2}{3}$ fraction's denominator to total 12 studs. Now the number of studs on each side is equivalent to 12.

3. Ask students: how many bricks were added to each side?

(*Answer:* Two 2x2 bricks were added to the $\frac{1}{4}$ fraction's denominator and three 1x3 bricks were added to the $\frac{2}{3}$ fraction's denominator.)

$\frac{1}{4}$ side = 2 bricks added

$\frac{2}{3}$ side= 3 bricks added

4. To find the numerator, add the same number of bricks to the top of each fraction.

For the ¼ fraction, add two 1x1 bricks to the numerator.

Now the fraction is $3/12$, which is equivalent to $1/4$.

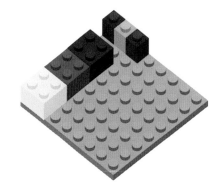

For the $2/3$ fraction, add three 1x2 bricks to the numerator.

Now the fraction is $8/12$, which is equivalent to $2/3$.

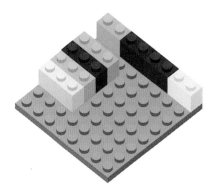

5. Write these equations: $1/4 + 2/3 = 11/12$ because $3/12 + 8/12 = 11/12$

Part 2: Show What You Know

1. Can you model the procedures and explain how to add $1/2 + 3/4$?

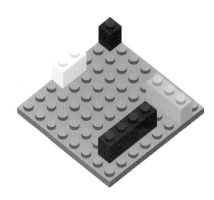

Students should first model the fractions. There are several ways to show this.

Students might notice that the $1/2$ fraction's denominator (1x2 brick) only needs one more to equal the $3/4$ fraction's denominator (1x4 brick). If not, they will add bricks to both denominators until they are equal, which is also acceptable but will result in larger denominators.

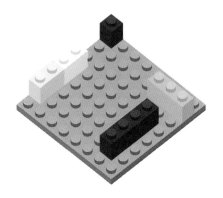

One 1x2 brick added to the ½ fraction's denominator makes 4 studs. The ¾ fraction's denominator does not need any additional bricks. Each side now has 4 studs.

½ fraction's denominator = 1 brick added

¾ fraction's denominator = 0 bricks added

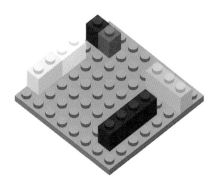

To show the addition, students should add the same number of bricks to the top (numerator) as they did to the bottom (denominator). Add one 1x1 brick to the ½ fraction's numerator, and no bricks to the ¾ fraction's numerator.

½ = ²⁄₄ and the ¾ fraction does not change
The solution is ²⁄₄ + ¾ = ⁵⁄₄.

Challenge: Can you determine what mixed number this is equal to?

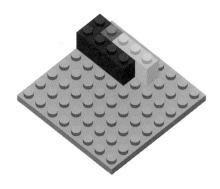

Students can do this by lining up the numerators in a row and comparing the number of studs to those in the denominator.

This shows that there is one whole and 1 extra stud. The stud represents ¼, since it takes 4 to make a whole in this case. So the mixed number is 1¼.

Note: Mixed numbers will be discussed in Chapter 9.

SUGGESTED BRICKS

Size	Number
1x1	10-20
1x2	8-10
1x3	15-20
2x2	15-20
2x8	4
	(or eight 2x4 bricks if necessary)

Note: Using a base plate will help keep the bricks in a uniform line. Use two small base plates or one large base plate for these activities.

SUBTRACTING FRACTIONS WITH UNLIKE DENOMINATORS

Students will learn/discover:
- How to subtract two fractions whose denominators are not the same

Why is this important?
Once students have a firm understanding of how to simplify and find equivalents for fractions, they are ready to subtract fractions of varying sizes. It is not possible to add two fractional values that have different wholes as a base, such as $1/3$ and $2/5$. Students need to be able to find a common base number in order to subtract fractional parts of a whole. A real-life example helps to show why this skill is important to learn: Sue ate $2/5$ of her container of yogurt and Tom ate $1/3$ of his yogurt. Who ate more yogurt? How much more did one eat than the other? In order to compare, students must first find a common denominator. Then the student must be able to subtract the two fractions to find the solution.

Brick Math journal:
After students build their models, have them draw the models on base plate paper and keep them in their Brick Math journals (see page 7 for instructions). Recording the models on paper after building with the LEGO® bricks helps to reinforce the concepts.

Part 1: Show Them How
Model subtracting two fractions: $\frac{2}{3}$ - $\frac{1}{4}$

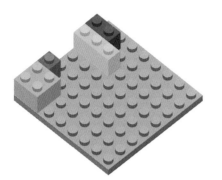

1. Place one 1x3 brick on the base plate with one 1x2 brick above it to model $\frac{2}{3}$. Place one 2x2 brick on the base plate with one 1x1 brick above it to model $\frac{1}{4}$.

Note: Use different colors of LEGO® bricks for numerator and denominator of both fractions to help demonstrate that they are different.

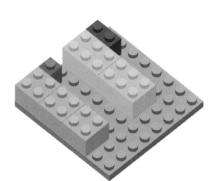

2. Since both denominators are different, add bricks to each denominator until both denominators have an equal number of studs.

Add three 1x3 bricks to the $\frac{2}{3}$ fraction's denominator to total 12 studs. Add two 2x2 bricks to the $\frac{1}{4}$ fraction's denominator to total 12 studs.

3. Ask students: how many bricks were added to each side?

Three 1x3 bricks were added to the $\frac{2}{3}$ fraction's denominator (top) and two 2x2 bricks were added to the $\frac{1}{4}$ fraction's denominator (bottom).

$\frac{2}{3}$ fraction = 3 bricks added

$\frac{1}{4}$ fraction = 2 bricks added

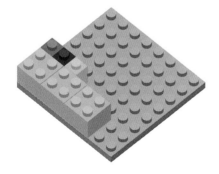

4. To find the numerator, add the same number of bricks to the top of each fraction.

For the $\frac{1}{4}$ fraction, add two 1x1 bricks to the numerator.

Now the fraction is $\frac{3}{12}$, which is equivalent to $\frac{1}{4}$.

For the $^2/_3$ fraction, add three 1x2 bricks to the numerator.

Now the fraction is $^8/_{12}$, which is equivalent to $^2/_3$.

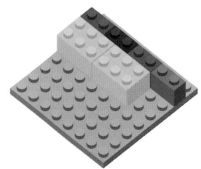

5. Model the subtraction problem $^8/_{12} - ^3/_{12}$.

Place the 1x3 brick (three studs) from the $^1/_4$ fraction's numerator on top of the 8 studs of the $^2/_3$ fraction's numerator.

Five studs in the numerator are not covered. This shows the solution to the problem: $^5/_{12}$.

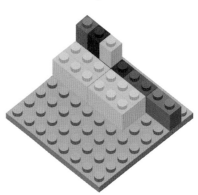

Part 2: Show What You Know

1. Can you model the procedures and explain how to subtract $^3/_4 - ^1/_2$?

Students should first model the fractions.

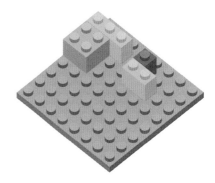

Students might notice that the $^1/_2$ fraction's denominator (1x2 brick) only needs one more brick to equal the $^3/_4$ fraction's denominator (2x2 brick). If not, they will add bricks until both are equal, which is also acceptable but will result in larger denominators.

One 1x2 brick added to the $^1/_2$ fraction's denominator makes 4 studs. The $^3/_4$ fraction's denominator does not need any additional bricks. Each side now has 4 studs.

$^1/_2$ fraction's denominator = 1 brick added
$^3/_4$ fraction's denominator = 0 bricks added

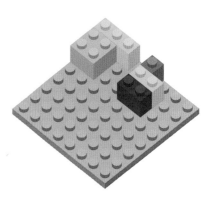

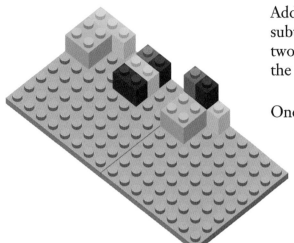

Add one 1x1 brick to the ½ fraction's denominator. Then subtract. To show the subtraction, students should place the two studs from the ½ fraction on top of the three studs of the ¾ fraction.

One stud is left uncovered. Therefore, the solution is ¼.

SUGGESTED BRICKS

Size	Number
1x1	10
1x2	8
1x3	8
2x4	8

Note: Using a base plate will help keep the bricks in a uniform line. One base plate is suggested for these activities.

MIXED NUMBERS

Students will learn/discover:
- to identify and model mixed numbers
- to understand the parts of a mixed number

Why is this important?

Mixed numbers (also called mixed fractions) show that a whole and a part of another whole can be written as one amount. For example, two bottles of juice and a half of another one would be written as the mixed fraction $2\frac{1}{2}$. This topic is important as students learn to add parts of various fractions together to create wholes with leftover fractional parts.

Brick Math journal:

After students build their models, have them draw the models on base plate paper and keep them in their Brick Math journals (see page 7 for instructions). Recording the models on paper after building with the LEGO® bricks helps to reinforce the concepts.

Part 1: Show Them How
Model the mixed number: 1¼

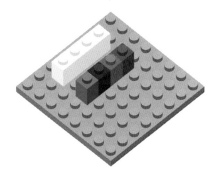

1. Place a 1x4 brick on the base plate and tell students it represents one whole.

Ask students: based upon what we have already learned about wholes, what does the whole represent? (*Answer:* 4 parts).

Place four 1x1 bricks next to the 1x4 brick and explain that the 1x1 bricks represent fourths.

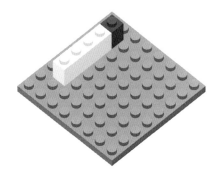

2. On another base plate, place a 1x4 brick and one 1x1 brick above it. Explain that it shows $1\frac{1}{4}$.

Use the terminology: A number like $1\frac{1}{4}$ is called a *mixed number*. It consists of an integer (the 1) and a fraction (the $\frac{1}{4}$).

Part 2: Show What You Know

1. Can you model the mixed number $2\frac{2}{4}$?

Ask students to explain their models.

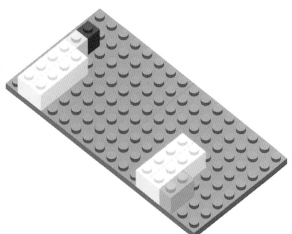

(*Answer:* Students can model the problem in several ways. Two are shown here. Both are correct, but model B (right) is the simpler form, which is preferable.

Model A (left) shows that the student understands the parts of the mixed number, but Model B (right) indicates that students understand that one 1x2 brick represents $\frac{1}{2}$, because $\frac{2}{4} = \frac{1}{2}$.)

2. Can you model other mixed number fractions?

Explain your models.

$1\frac{3}{8}$

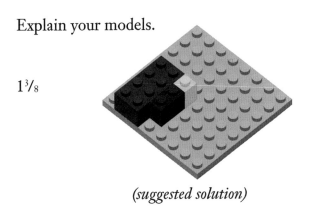

(suggested solution)

$3\frac{4}{10}$

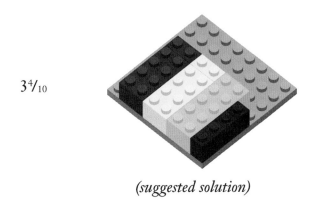

(suggested solution)

SUGGESTED BRICKS

Have a wide assortment of bricks available for the challenge.

Note: Using a base plate will help keep the bricks in a uniform line. One base plate is suggested for these activities.

FRACTION PUZZLE CHALLENGE

This activity brings together all the knowledge gained from the previous nine chapters. It assumes that students can find fractional parts of a whole, add fractions with like and unlike denominators, and find equivalents. Most of the skills in this activity are appropriate for grades 3-6.

Brick Math journal:
After students build their models, have them draw the models on base plate paper and keep them in their Brick Math journals (see page 7 for instructions). Recording the models on paper after building with the LEGO® bricks helps to reinforce the concepts.

1. Have students create this exact model and examine it carefully. Be sure to use the same colors of bricks as in the model shown, since some answers reference the colors.

These bricks are used to create this model:
(Total of 10 bricks)

Size	Number	Color
1x2	1	light green
2x2	3	dark green, orange, yellow
2x3	1	light green
(or two 1x3 bricks)		
2x4	4	yellow, blue, 2 red
2x6	1	dark green

2. Ask students to gather the following data from their models:

a. What is the fractional part of the whole for each color?

Yellow _____ (*Answer:* $^{12}/_{64}$ or $^{3}/_{16}$)

Red _____ (*Answer:* $^{16}/_{64}$ or $^{1}/_{4}$)

Blue_____(*Answer:* $^{8}/_{64}$ or $^{1}/_{8}$)

Dark green _____ (*Answer:* $^{16}/_{64}$ or $^{1}/_{4}$)

Light green _____ (*Answer:* $^{8}/_{64}$ or $^{1}/_{8}$)

Orange _____ (*Answer:* $^{4}/_{64}$ or $^{1}/_{16}$)

b. Which colors represent equivalent amounts?

(*Answer:* Blue and light green are equivalent;
Red and dark green are equivalent)

c. Blue + light green = _____(*Answer:* $^{1}/_{4}$)

d. Dark green + yellow = _____ (*Answer:* $^{7}/_{16}$)

e. What combinations of colors are equivalent to ½ of
the base plate?

(*Possible Answers:*
Red + dark green
Yellow + blue + orange + light green
Light green + blue + dark green
Yellow + orange + red
Yellow + orange + dark green
Red + blue + light green)

Additional Challenge: Have each student create a new puzzle
and write questions for a partner to solve. Be sure each student
provides an answer key to check the partner's answers.

APPENDIX

Base Plate Paper

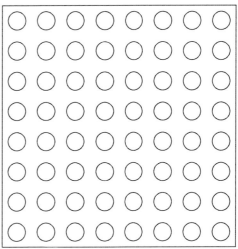

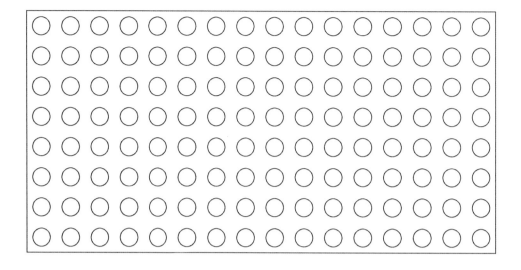

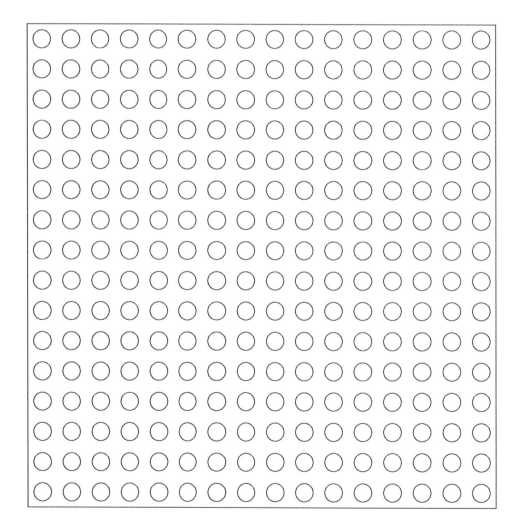

Printed in Great Britain
by Amazon